Contents

Name_______________________________ Class_____________________ Date___________

Concept Review

Section: Sound

1. **Explain** why the speed of sound changes if the temperature of the medium changes.

____

____

____

2. **Explain** what factors affect the loudness of a sound.

____

____

3. **Describe** how to change the pitch of a note played on a stringed instrument.

____

____

____

4. **Describe** the difference between infrasound and ultrasound, and provide one example of how each is used.

____

____

5. **Describe** how sound waves travel from the air through the ear.

____

____

____

6. **Determine** the following distances based on the sonar data given. A ship sends a sound pulse downward and receives the reflected sound 2.50 s later. (**Hint:** Use the formula $d = vt$. Assume the speed of sound in sea water is 1,540 m/s.)

   a. the total distance traveled by the sound pulse

____

   b. the depth of the water

____

Name _________________________________ Class _________________ Date ______________

Concept Review

Section: The Nature of Light

1. **Explain** what is meant by the "dual nature" of light.

   ____

2. **Indicate** whether the following behaviors of light can best be described in terms of the particle model or the wave model.

   _________________________ a. Light can produce interference patterns.

   _________________________ b. Light can travel in a vacuum.

   _________________________ c. Dim blue light can knock electrons off a metal plate, whereas bright red light cannot.

3. **Determine** which band of the electromagnetic spectrum has each of the following:

   _________________________ a. the longest wavelength

   _________________________ b. the highest frequency

   _________________________ c. the greatest energy

   _________________________ d. the least energy

4. **Explain** why intensity decreases as distance from the light source increases.

   ____

   ____

   ____

5. **List** the regions of the electromagnetic spectrum in order of increasing energy.

   ____

   ____

6. **Describe** one use of X rays and one use of gamma rays in medicine.

   ____

   ____

   ____

Skills Worksheet

Concept Review

Section: Reflection and Color

1. **Contrast** the reflection of light from rough surfaces with that from smooth surfaces. Why do rough surfaces cause diffuse reflection?

____

____

____

2. **Use** the law of reflection to draw a sketch showing the incoming and reflected light rays when light shines on a mirror at an angle of 30° to the normal. Label the angle of incidence and the angle of reflection.

3. **Describe** how the eye would perceive an image created by a concave mirror and by a convex mirror.

____

____

4. **Draw** a diagram that shows how a flat mirror forms a virtual image.

5. **Indicate** what color a yellow cloth would appear to be if it were illuminated with

_____________________ a. sunlight.

_____________________ b. yellow light.

_____________________ c. blue light.

Skills Worksheet

Concept Review

Section: Refraction, Lenses, and Prisms

1. **Indicate** the direction that light is bent in each of the following situations.

   a. Light passes from air to water.

   ____

   b. Light passes from water to air.

   ____

2. **Explain** what causes a mirage.

   ____

   ____

3. **List** the colors in the visible spectrum in order from longest to shortest wavelength.

   ____

   ____

   ____

4. **Name** the type of lens that creates either virtual or real images, and state the direction that light rays are bent by this type of lens.

   ____

   ____

5. **Describe** the function of each of the following parts of the eye.

   a. cornea

   ____

   b. pupil

   ____

   c. lens

   ____

6. **Explain** why a prism can separate white light.

   ____

   ____

Skills Worksheet

Science Skills

Angles and Degrees

Wherever two straight lines meet, they form an *angle*. The size of an angle is measured in units called *degrees*. Degrees are often abbreviated with the symbol °. For example, 75 degrees can be written 75°.

The angle between two perpendicular lines is equal to 90°. A 90° angle is called a *right angle*. A triangle with a right angle is called a *right triangle*.

The three angles of a triangle always add up to 180°. In a right triangle, one angle equals 90°, so the other two angles must add up to 90° (for a total of 180°). If you know two angles in a triangle, you can always determine the third angle.

PROBLEM

If a right triangle has one 35° angle, what is the other non-right angle?

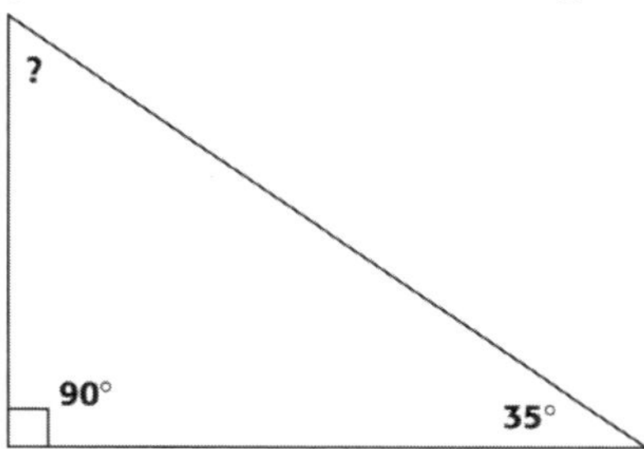

SOLUTION

Step 1: Write down the two given angles, and add them together. Because the triangle is a right triangle, one angle must equal 90°. We also know that one of the other angles equals 35°.

$$90° + 35° = 125°$$

Step 2: Subtract the sum of the two known angles from 180°. Because all three angles must add up to 180°, subtracting the sum of the two known angles from 180° will give the value of the third, unknown angle.

$$180° - 125° = 55°$$

PRACTICE

Work out the practice problems on a separate sheet of paper, and write your answers in the spaces provided.

1. If one angle of a right triangle equals 60°, what is the other non-right angle?

____

2. If two angles of a triangle equal 65° each, what must the third angle be?

____

3. Two angles of a triangle equal 20° and 70°. Is this a right triangle?

____

4. What is the sum of the four angles in a rectangle? (**Hint:** A rectangle can be divided into two triangles.)

____

Science Skills

The Area of a Circle

Suppose you are trying to decide whether to order a pizza that is 30.0 cm across (about 12 in.) or a pizza that is 25.0 cm across (about 10 in.). How much bigger is the 30.0 cm pizza? To find out, you need to know the area of each pizza.

To calculate the area of any circle, you can use the following equation:

$$Area\ of\ a\ circle = \pi \times r^2$$

In this equation, the symbol π stands for pi, a constant roughly equal to 3.14, and r^2 is the radius squared. The radius is the distance from the outside of the circle to the center. Sometimes, you may know the diameter instead of the radius. The diameter is any line drawn across the circle through the center. The radius is always equal to one-half the diameter.

PROBLEM

Determine the area of a pizza that is 30.0 cm across.

SOLUTION

Step 1: Determine the radius. The pizza is 30.0 cm across, so 30.0 cm is the diameter of the circle. Divide the diameter in half to find the radius:

$$radius = \frac{diameter}{2} = \frac{30.0\ cm}{2} = 15.0\ cm$$

Step 2: Substitute values. Substitute the radius into the area equation.

$$Area\ of\ a\ circle = \pi \times r^2 = \pi \times (15.0\ cm)^2$$

Step 3: Multiply known values. Multiply all of the terms on the right side of the equation. For π, use the π button on your calculator.

$$\pi \times (15.0\ cm)^2 = 706.858\ cm^2$$

Step 4: Round to the correct number of significant figures. Round to the lowest number of significant figures in the calculation, which is 3.

$$Area = 707\ cm^2$$

PRACTICE

Work out the practice problems on a separate sheet of paper, and write your answers in the spaces provided.

1. Find the area of a pizza that is 25.0 cm across.

   ____

2. Determine the area of a round swimming pool with a radius of 8.0 m.

   ____

Skills Worksheet

Science Skills

Equations Involving a Constant

When you are solving a word problem in science, most of the values you need will be given. Any value that is not given in the problem is probably a constant, which always has the same value. A few of the constants you will encounter in your study of science are given in the following table.

Constant	Approximate Value
g (free-fall acceleration at sea level)	9.8 m/s^2
G (universal gravitation)	$6.7 \times 10^{-11} \text{ N} \cdot \text{m}^2/\text{kg}^2$
c (speed of light in a vacuum)	$3.0 \times 10^8 \text{ m/s}$

PROBLEM

Red light has a frequency (f) of 4.3×10^{14} Hz. Use the equation $c = f \times \lambda$ to find the wavelength (λ) of red light.

SOLUTION

Step 1: Rearrange the equation if necessary. In this case, you need to find the wavelength, so divide both sides of the equation by f to isolate λ.

$$\lambda = \frac{c}{f}$$

Step 2: Look in a table of constants for any values that are not given. Look up the value for c, the speed of light, in the table on this page.

Step 3: Substitute values into the equation, and solve. The unit Hz (hertz) is equivalent to 1/s, so the units of s cancel, leaving units of m.

$$\lambda = \frac{c}{f} = \frac{3.0 \times 10^8 \text{ m/s}}{4.3 \times 10^{14} \text{ Hz}} = 6.97 \times 10^{-7} \text{ m}$$

Step 4: Round to the correct number of significant figures. In this case, round the answer to two significant figures: 7.0×10^{-7} m.

PRACTICE

1. Use the equation in the example to find the wavelength of violet light with a frequency of 7.5×10^{14} Hz.

____

2. Find the weight *(weight = mg)* of a 17 kg crate of books.

____

Skills Worksheet

Cross-Disciplinary

The Refracting Telescope at Yerkes

Read the following paragraphs, and complete the exercises below.

WORLD'S LARGEST REFRACTING TELESCOPE

The 40-inch telescope at Yerkes Observatory, in Williams Bay, Wisconsin, is the largest refracting telescope in the world. This telescope uses glass lenses—each 40 inches across—to focus light.

The 40-inch telescope at Yerkes is enclosed by a dome that measures 90 ft across. The telescope's tube is 63 ft long, and it rests on a movable floor that hangs from four strong cables. When the telescope is moved to aim at different locations in space, the height of its eyepiece changes. The movable floor raises astronomers to reach the eyepiece.

DISCOVERIES AT YERKES

The telescope at Yerkes has been the site of many discoveries. For example, data collected by the 40-inch telescope between 1937 and 1951 showed that our galaxy, the Milky Way, has a spiral shape. The fifth moon of Uranus and the second moon of Neptune were also discovered from Yerkes.

Current projects at Yerkes include a study of the motion of stars. Stars move so slowly that it is difficult to detect their motion, even over a long period of time. To track a star's movement, astronomers need data taken over many years with the same instrument. One astronomer at Yerkes compares pictures taken with the 40-inch telescope before 1910 with new photographs. With this method, the astronomer can obtain extremely precise measurements of star movement.

EXERCISES

1. What is unique about the 40-inch telescope at Yerkes Observatory?

   ____

   ____

2. Why is there a movable floor beneath the refracting telescope at Yerkes?

   ____

   ____

3. Explain why tracking the movements of stars requires studying images taken over a period of years.

   ____

   ____

 REAL WORLD APPLICATIONS

Cross-Disciplinary

How Does Sunscreen Work?

Read the following paragraphs, and complete the exercises below.

Have you ever wondered how the sun's rays can cause sunburns and suntans? When too much ultraviolet (UV) radiation from the sun strikes your skin, small vessels in the skin expand. These vessels contain groups of red and white blood cells, which produce the red coloring that is known as a sunburn.

The body's defense against sunburn is to release a chemical pigment, called melanin, to the surface of the skin. Melanin absorbs UV radiation from the sun, preventing it from penetrating the skin tissue. The melanin also causes the dark coloring that we call a suntan. When the melanin cells die, the suntan fades.

SUNSCREENS WORK LIKE MELANIN

Like melanin, sunscreens contain pigments that absorb UV radiation from the sun to keep it from penetrating the skin. The SPF of a sunscreen tells how much radiation it absorbs. For example, SPF 10 absorbs all but 1/10 of the sun's ultraviolet radiation, while SPF 40 absorbs all but 1/40 of it.

SPF ratings are only approximate because they depend on how much sunscreen you apply. People with light complexions have less melanin pigment. As a result, they must use higher levels of SPF to be protected from the sun's ultraviolet rays.

EXERCISES

1. What causes sunburns and suntans?

____

____

2. How are the pigments in sunscreen similar to melanin?

____

____

3. Suppose you are wearing sunscreen with SPF 20. How much of the sun's UV radiation will be absorbed by the sunscreen? How much will penetrate your skin?

____

____

Assessment

Pretest

Sound and Light

______ 1. Sound waves are examples of
   a. longitudinal waves.
   b. transverse waves.
   c. mechanical waves.
   d. circular waves.

______ 2. In a vacuum, the speed of light is
   a. 9.8 m/s2.
   b. $3 \times 108 \text{ m/s}$.
   c. $1.97 \times 108 \text{ m/s}$.
   d. $2.05 \times 108 \text{ m/s}$.

______ 3. Interference between two waves can produce a wave that has a
   a. larger amplitude.
   b. smaller amplitude.
   c. larger amplitude or smaller amplitude.
   d. None of the above

______ 4. Which of the following involves waves changing direction?
   a. refraction   c. reflection
   b. diffraction   d. All of the above

5. Draw a longitudinal wave and a transverse wave in the space below. On each wave, indicate the wavelength. On the transverse wave, indicate the amplitude.

6. How can you determine the frequency of a wave?

____

____

____

Assessment

Quiz

Section: Sound

In the space provided, write the letter of the term or phrase that best completes each statement or best answers each question.

______ 1. Which of these affects the speed of sound?
    a. loudness of the sound   c. temperature
    b. resonance   d. direction of the wave

______ 2. Which of these affects the loudness of a sound?
    a. frequency of the waves   c. pitch of the sound
    b. intensity of the waves   d. speed of the sound

______ 3. Which of these affects the pitch of a sound?
    a. loudness of the sound   c. intensity of the waves
    b. frequency of the waves   d. resonance

______ 4. Nerve fibers send an impulse to the brain when the waves reach the
    a. cochlea.   c. eardrum.
    b. hammer.   d. anvil.

______ 5. To hear, which of the following takes place?
    a. The ear senses vibrations in the air.
    b. The ear amplifies the vibrations.
    c. The ear transmits signals to the brain.
    d. All of the above

______ 6. What causes the sound of a guitar to get louder as it is played?
    a. infrasound   c. resonance
    b. pitch   d. ultrasound

______ 7. Which of these is used to measure distance?
    a. infrasound   c. antisound
    b. ultrasound   d. All of the above

______ 8. Sonar uses
    a. the dispersion of sound waves.   c. ultrasound.
    b. low-frequency sound waves.   d. All of the above

______ 9. Which device uses sound waves to view organs inside the body?
    a. X-ray machine   c. radar
    b. sonogram   d. MRI

______ 10. Because its molecules are far apart, sound waves will travel slowly through
    a. cold air.   c. iron.
    b. water.   d. copper.

Assessment

Quiz

Section: The Nature of Light

In the space provided, write the letter of the term or phrase that best completes each statement or best answers each question.

______ 1. Light demonstrates wave characteristics when it is
   a. reflected.
   b. refracted.
   c. diffracted.
   d. All of the above

______ 2. Light demonstrates particle characteristics when it
   a. knocks electrons off a metal surface.
   b. passes through a narrow opening.
   c. forms standing waves.
   d. All of the above

______ 3. As the frequency of light waves increases,
   a. the energy increases.
   b. the energy decreases.
   c. the energy stays the same.
   d. the wavelength increases.

______ 4. The rate at which energy flows through a given space describes light
   a. resonance.
   b. pitch.
   c. intensity.
   d. interference.

In the space provided, write the letter of the term or phrase that best matches each description.

______ 5. electromagnetic waves that give off heat

______ 6. electromagnetic waves that can carry telecommunication signals and cook food

______ 7. electromagnetic waves with high energy that can cause sunburn

______ 8. electromagnetic waves used in medicine to kill cancer cells and view bones

______ 9. electromagnetic waves used for television, AM, and FM radio signals

______ 10. electromagnetic waves used to determine the location of objects in a room

a. ultraviolet rays
b. infrared rays
c. visible light
d. radio waves
e. microwaves
f. X rays and gamma rays

Assessment

Quiz

Section: Reflection and Color

In the space provided, write the letter of the term or phrase that best completes each statement or best answers each question.

_______ 1. Light rays reflecting off a smooth surface reflect
    a. in one new direction.   c. in random directions.
    b. in the initial direction.   d. None of the above

_______ 2. Light rays reflecting off a rough surface reflect
    a. diffusely.   c. at all angles.
    b. at a single angle.   d. None of the above

_______ 3. The law of reflection says the angle of incidence is ____________ the angle of reflection.
    a. greater than   c. equal to
    b. less than   d. None of the above

_______ 4. Reflection of light into random directions is called
    a. diffuse reflection.   c. the angle of reflection.
    b. the law of reflection.   d. the reflective ray.

_______ 5. A virtual image
    a. appears in front of a mirror.   c. diffuses light.
    b. appears behind a mirror.   d. reflects light.

_______ 6. An apple appears red to your eye because it
    a. emits red light.   c. reflects red light.
    b. absorbs red light.   d. distorts red light.

_______ 7. When you look in a convex mirror, the image appears
    a. smaller than it really is.   c. larger than it really is.
    b. lighter than it really is.   d. darker than it really is.

_______ 8. In addition to reflecting some light, every object
    a. transmits radio waves.   c. produces vibrations.
    b. absorbs some light.   d. None of the above

_______ 9. A real image can be formed using a
    a. concave mirror.   c. flat mirror.
    b. convex mirror.   d. All of the above

_______ 10. When all three additive primary colors are mixed, you see the color
    a. black.   c. cyan.
    b. magenta.   d. white.

Quiz

Section: Refraction, Lenses, and Prisms

In the space provided, write the letter of the term or phrase that best completes each statement or best answers each question.

______ 1. When light moves from a material in which its speed is higher to one in which its speed is lower, the ray is
   a. bent away from the normal.
   b. bent toward the normal.
   c. converged.
   d. diverged.

______ 2. A mirage is caused by
   a. dispersing light.
   b. converging light.
   c. diverging light.
   d. refracting light.

______ 3. Light entering the eye is focused to produce an image on the
   a. iris.
   b. cornea.
   c. retina.
   d. optic nerve.

______ 4. A prism separates white light into different colors through a process called
   a. dispersion.
   b. magnification.
   c. convergence.
   d. divergence.

In the space provided, write the letter of the term or phrase that best matches each description.

______ 5. the increase of an object's apparent size by using lenses or mirrors

______ 6. consists of two or more plane surfaces of a transparent solid at an angle with each other

______ 7. a transparent object that refracts light waves such that they converge or diverge to create an image

______ 8. a lens that bends light waves inward

______ 9. bending of light when crossing between mediums

______ 10. a lens refracting light waves so that they bend outward

a. refraction
b. prism
c. convergent lens
d. divergent lens
e. magnification
f. lens

Assessment **TEST A**

Chapter Test

Sound and Light

In the space provided, write the letter of the term or phrase that best completes each statement or best answers each question.

_______ 1. The intensity of a sound wave depends on
   a. the wave's loudness.
   b. the wave's amplitude.
   c. pitch.
   d. frequency.

_______ 2. Resonance amplifies the sound of a guitar when the
   a. the guitar string and guitar body are both vibrating at the same frequency.
   b. intensity of a sound decreases over time.
   c. pitch of a note is compared with a pure tone.
   d. vibration of a string or column of air causes a standing wave at a natural frequency.

_______ 3. The particle model of light explains how light can
   a. travel through empty space without a medium.
   b. refract when it passes through a lens.
   c. be reflected off a mirror.
   d. diffract when it passes through a normal opening.

_______ 4. The amount of energy in a photon of light is proportional to the
   a. medium through which it travels.
   b. shape of the light wave it creates.
   c. speed of the corresponding light wave.
   d. frequency of the corresponding light wave.

_______ 5. When light rays reflect off a rough surface, they
   a. scatter in many different directions.
   b. converge toward the normal.
   c. diverge away from the normal.
   d. decrease their speed and change their angle.

_______ 6. When you look at an orange, you see the color orange because the fruit
   a. reflects orange light and absorbs other colors.
   b. absorbs orange light and reflects other colors.
   c. reflects red and yellow light only.
   d. absorbs red and yellow light only.

Chapter Test *continued*

_______ 7. When light moves from a material in which its speed is higher to a
material in which its speed is lower, it is
a. bent toward the normal.   c. reflected off the boundary.
b. bent away from the normal.   d. changed into a virtual image.

_______ 8. Which of the following uses radio waves to find an object's location?
a. sonar   c. sonogram
b. ultrasound   d. radar

_______ 9. Which structure within the eye is responsible for the largest percentage
of refraction of light?
a. retina   c. lens
b. cornea   d. iris

_______ 10. White light separates into different colors when it passes through a
prism because of
a. differences in wave speed.   c. reflection.
b. total internal dispersion.   d. droplets in the air.

**Read each statement, and write in the blank the word or words that best completes
the statement.**

11. The greater the _______________________ of a sound wave, the louder the

sound.

12. Reflected sound waves make it possible for _______________________

devices to measure distance.

13. The three small bones in your middle ear are the _______________________,

the _______________________, and the _______________________.

14. In the particle model of light, individual "packets" of light are called

_______________________.

15. Radio waves, microwaves, and infrared waves make up a part of the

_______________________.

16. The theoretical line perpendicular to the surface where a light ray hits a mirror

is called the _______________________.

17. When you look into a flat mirror, you see a(n) _______________________

image.

Chapter Test *continued*

18. When sunlight is dispersed and reflected by water droplets, a(n)

    __________________ forms.

19. If light moves from a material in which its speed is lower to one in which its

    speed is higher, the ray is bent away from the __________________ .

Read the question, and write your response in the space provided.

20. What does it mean to say that light can be modeled as both a wave and
    particle? Explain the implications of this for scientific theory.

    ____

    ____

    ____

    ____

    ____

    ____

Chapter Test

Sound and Light

In the space provided, write the letter of the term or phrase that best completes each statement or best answers each question.

_______ 1. Most musical instruments produce sound through the vibration of
   a. membranes.
   b. air columns.
   c. strings.
   d. All of the above

_______ 2. Sonar and ultrasound technology depend on the ability of waves to
   a. reflect sound.
   b. amplify sound.
   c. disperse sound.
   d. dissipate sound.

_______ 3. Which property of light is not explained by the wave model of light?
   a. Light produces interference patterns.
   b. Light is diffracted when it passes through a narrow opening.
   c. Blue light can knock electrons off a plate but red light cannot.
   d. Light reflects when it meets a mirror, but it refracts when it passes
      through a lens.

_______ 4. The rate at which light energy flows through a given area of space is
   referred to as its
   a. speed.   c. intensity.
   b. wavelength.   d. energy.

_______ 5. The law of reflection states that when light rays reflect off a surface,
   the angle of incidence
   a. is one-half the angle of reflection.
   b. equals the angle of reflection.
   c. is twice the angle of reflection.
   d. equals the angle of refraction.

_______ 6. An image that results from the apparent path of light rays is called a(n)
   a. distorted image.   c. objective image.
   b. real image.   d. virtual image.

Chapter Test *continued*

_______ 7. If you looked at a red tulip with green leaves under green light, you
would see a
   a. green tulip with green leaves.
   b. black tulip with green leaves.
   c. yellow tulip with black leaves.
   d. black tulip with black leaves.

_______ 8. A virtual image caused by reflection of light in the atmosphere is
called a
   a. prism.   c. mirror.
   b. lens.   d. mirage.

_______ 9. Which statement about a diverging lens is correct?
   a. It bends light inward and can create either a virtual or a real image.
   b. It bends light inward and can only create a real image.
   c. It bends light outward and can create either a virtual or a real image.
   d. It bends light outward and can only create a virtual image.

_______ 10. The effect in which white light separates into different colors is called
   a. magnification.   c. reflection.
   b. refraction.   d. dispersion.

**Read each statement, and write in the blank the word or words that best completes
the statement.**

11. Sound will travel more quickly through air if the _______________________________

   of the air increases.

12. In your inner ear, different parts of the _______________________________ vibrate at

   different natural frequencies.

13. Sound waves with frequencies higher than 20,000 Hz are referred to as

   _______________________________ waves.

14. The two most common models of light describe it as a wave or as a stream of

   _______________________________.

15. All possible kinds of light, at all energies, frequencies, and wavelengths, make

   up the _______________________________.

16. Reflection of light in random directions is called _______________________________.

17. A mirror that curves outward is a(n) _______________________________ mirror.

18. An object looks red to your eye if it _____________________ red light and

_____________________ all other colors.

19. Light waves _____________________ when they pass from one transparent

medium to another.

Read the statement, and write your response in the space provided.

20. Describe two different forms of electromagnetic waves and explain their uses
in technology.

____

____

____

____

____

Standardized Test Practice

Read the passage below and answer the questions that follow.

This advertisement was seen in the newspaper.

ON SALE THIS WEEKEND ONLY: the *Mega-Magnificent Microwave Oven* for only $599! This must-have microwave is designed for every kitchen—from contemporary to modern. You will marvel at how easily and quickly your meals are prepared. In fact, with the *Mega-Magnificent Microwave Oven,* complete meal preparation can be reduced to no more than 10 minutes!! Whether it's meat, poultry, fish, vegetables, or even desserts, the *Mega-Magnificent Microwave Oven* cooks meals that are aromatic, savory, and nutritious.

When you purchase the *Mega-Magnificent Microwave,* you'll receive—at no extra charge—the roasting, casserole, bacon, and sautéing dishes.

But hurry in today! Our supply is limited, and future deliveries can't be guaranteed. So, NO RAIN CHECKS WILL BE ISSUED.

In the space provided, write the letter of the term or phrase that best completes each statement or best answers each question.

_______ 1. What is the main purpose of the passage?
   A to educate consumers about microwave cooking
   B to promote microwave oven use
   C to sell *Mega-Magnificent Microwave Ovens*
   D to identify the advantages of microwave ovens

_______ 2. Which of the following statements is a FACT?
   A The *Mega-Magnificent Microwave Oven* is a must-have kitchen appliance.
   B The *Mega-Magnificent Microwave Oven* is on sale this weekend.
   C Meal preparation is guaranteed to take no longer than 10 minutes.
   D There is no easier way to prepare meals for your family.

_______ 3. What is the likely reason for including the following sentence: "Our supply is limited, and future deliveries can't be guaranteed"?
   A The advertiser thought you might want to know.
   B Truth-in-advertising laws require this disclosure.
   C The advertiser wanted to inform the readers about supply.
   D The advertiser wanted to increase the sense of urgency.

_______ 4. You can tell from the passage that
   A this microwave oven has some useful features and accessories.
   B microwave ovens are in short supply.
   C there is no better microwave oven.
   D this microwave oven is easiest to operate.

Inquiry Lab   **DATASHEET**

Colors in White Light

MATERIALS

For each group

- light bulb, fluorescent
- light bulb, incandescent
- spectroscope

PROCEDURE

Your teacher will give you a spectroscope or instructions for making one. Turn on an incandescent light bulb. Look at the light bulb through your spectroscope. Write a description of what you see. Then, repeat the procedure with a fluorescent light bulb. Again, describe what you see.

QUESTIONS TO GET YOU STARTED

1. Compare your observations of the incandescent light bulb and your observations of the fluorescent light bulb.

   ____

   ____

   ____

2. Both kinds of bulbs produce white light. What did you learn about white light by using the spectroscope?

   ____

   ____

   ____

3. Light from a flame is yellowish but is similar to white light. What do you think you would see if you used a spectroscope to look at light from a flame?

   ____

   ____

   ____

Quick Lab   **DATASHEET**

Sound in Different Mediums

MATERIALS

For each group

- spoon, or other utensil
- string, 1–2 m

PROCEDURE

1. Tie a spoon or other utensil to the middle of a 1 to 2 m length of string.
2. Wrap the loose ends of the string around your index fingers, and place your fingers against your ears.
3. Swing the spoon so that it strikes a tabletop.

ANALYSIS

1. Compare the volume and quality of the sound received through the string with the volume and quality of the sound received through the air.

   ____

   ____

2. Does sound travel better through the string or through the air? Explain.

   ____

   ____

   ____

3. Explain your results.

   ____

   ____

   ____

Quick Lab **DATASHEET**

Frequency and Pitch

MATERIALS

For each group

- desk
- ruler, flexible metal or plastic

PROCEDURE

1. Hold one end of a flexible metal or plastic ruler on a desk with about half of the ruler hanging off the edge. Bend the free end of the ruler, and then release it. Can you hear a sound?

   ____

2. Try changing the position of the ruler so that less hangs over the edge. How does the change in position change the sound produced?

   ____

   ____

   ____

   ____

   ____

Inquiry Lab   **DATASHEET**

How Can You Amplify the Sound of a Tuning Fork?

MATERIALS

For each group

- block, rubber
- objects, wood or metal

- tuning forks, 2 of the same frequency and at least 1 of a different frequency

PROCEDURE

1. Activate a tuning fork by striking the tongs of the fork against a rubber block.

2. Touch the base of the tuning fork to various wood or metal objects. Listen for any changes in the sound of the tuning fork.

3. Activate the fork again. Then, try touching the base of the tuning fork to the bases of other tuning forks. Make sure that the tines of the forks are free to vibrate and are not touching anything.

4. If you find two tuning forks that resonate with each other, try activating one and holding it near the tines of the other one.

ANALYSIS

1. What are some characteristics of the objects that helped to amplify the sound of the tuning fork in step 2?

____

____

2. In step 3, could you make another tuning fork start vibrating by touching it with the base of your tuning fork?

____

3. In step 4, could you make another tuning fork vibrate without touching it with your tuning fork?

____

4. What is the relationship between the frequencies of tuning forks that resonate with each other in steps 3 and 4?

____

____

Quick Lab   **DATASHEET**

Echoes and Distance

MATERIALS

For each group

- stopwatch

PROCEDURE

1. Stand outside in front of a large wall, and clap your hands. Do you hear an echo?

   ____

   ____

2. Use a stopwatch to measure the time that passes between the time you clap your hands and the time you hear the echo.

   ____

   ____

3. The speed of sound in air is about 340 m/s. Use the approximate time from step 2 and the speed equation to estimate the distance to the wall.

   ____

   ____

   ____

Quick Lab   **DATASHEET**

Curved Mirror

MATERIALS

For each group

- pencil, short

- spoon, large, stainless steel

PROCEDURE

1. Observe the reflection of a short pencil in the inner part of a large, stainless steel spoon.

2. Slowly move the spoon closer to the pencil. Note any changes in the appearance of the pencil's reflection.

3. Repeat steps 1 and 2 by using the other side of the spoon as the mirror.

ANALYSIS

1. Which side of the spoon is a concave mirror? Which side is a convex mirror?

____

____

____

2. What differences in the reflected images did you observe?

____

____

____

3. How does distance affect an object's image in concave and convex mirrors?

____

____

____

Filtering Light

MATERIALS

For each group

- filters, colored (4)

PROCEDURE

1. Look through one of four colored filters—red, blue, yellow, or green—at an object across the room. Describe the object's color.

   ____

   ____

2. Repeat step 1 with each one of the filters.

   ____

   ____

   ____

   ____

   ____

3. Look at the same object through two filters placed together. Describe the object's color. Why does the object's color differ from the color when one filter is used?

   ____

   ____

   ____

4. Place the red, blue, and yellow filters together, and look at the same object. Describe what you see. Why do you see this?

   ____

   ____

   ____

   ____

Quick Lab   **DATASHEET**

Water Prism

MATERIALS

For each group

- dish, shallow
- mirror, flat
- paper, white, 1 sheet
- sunlight
- water

PROCEDURE

1. Fill a shallow dish with about 3 cm of water.

2. Place a flat mirror in the tray so that about half of the mirror is under the water. The mirror should be placed at an angle and should lean against the side of the dish.

3. Use a white piece of paper to capture the sunlight that is reflected by the mirror. Change the distance of the paper from the mirror and the angle of the paper until you see colors in the reflection. What colors do you see?

____

____

____

4. Why are the colors of the reflected sunlight separated by the water?

____

____

____

Application Lab **DATASHEET A**

Lenses and Images

As an optical engineer for a camera company, you have been given a lens for which your job is to figure out the focal length. Based on the specifications you obtain by doing an experiment, a new model of camera will be designed that uses that lens.

WHAT YOU'LL DO

Observe images formed by a convex lens.

Measure the distance of objects and images from the lens.

Analyze your results to determine the focal length of the lens.

WHAT YOU'LL NEED

- cardboard screen, 10 cm × 20 cm
- convex lens, 10 cm to 15 cm focal length
- lens holder
- light box with light bulb

- meterstick
- ruler, metric
- screen holder
- supports for meterstick

PROCEDURE

Preparing for Your Experiment

1. What you will do:
   - The shape of a lens determines the size, position, and types of images that it may form.
   - When parallel rays of light from a distant object pass through a converging lens, they come together, or converge, to form an image at a point called the *focal point*.
   - The distance from the focal point to the lens is called the *focal length*.
   - In this experiment, you will find the focal length of a lens.
   - You will then check this value by forming images, measuring distances, and using the lens formula,

$$\frac{1}{d_o} + \frac{1}{d_i} = \frac{1}{f}$$

where d_o = *object distance,*

d_i = *image distance,* and

f = *focal length*

Lenses and Images *continued*

2. Use the following table to record your data.

DATA TABLE: OBJECTS, LENSES, AND IMAGES

Focal length of lens, f: ______cm	Object distance, d_o (cm)	Image distance, d_i (cm)	$\dfrac{1}{d_o}$	$\dfrac{1}{d_i}$	$\dfrac{1}{d_o}+\dfrac{1}{d_i}$	$\dfrac{1}{f}$	Size of object (mm)	Size of image (mm)
Trial 1								
Trial 2								
Trial 3								

3. Set up the equipment as illustrated in the figure below.
   • Make sure the lens and screen are securely fastened to the meterstick.

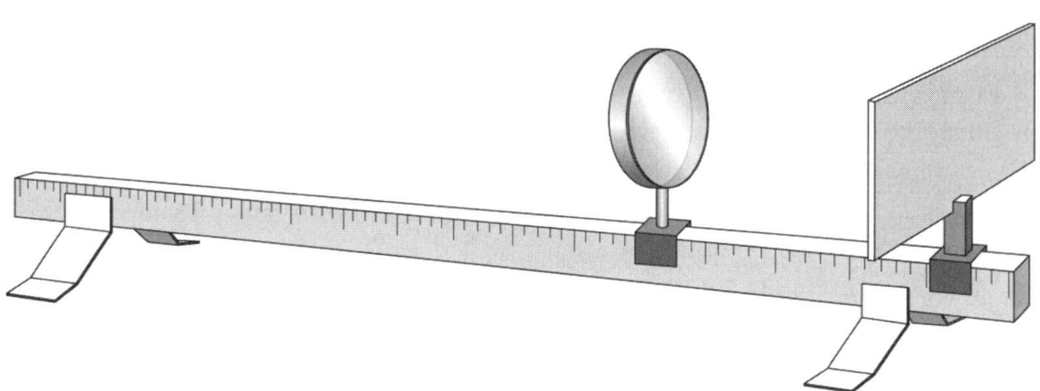

Determining Focal Length

4. Find the focal length.
   • Stand about 1 m from a window.
   • Point the meterstick at a tree, parked car, or similar object.
   • Look through the lens.
   • Slide the screen holder along the meterstick until a clear image of the distant object forms on the screen.
   • Measure the distance between the lens and the screen in centimeters.
   • This distance is very close to the focal length of the lens that you are using.
   • Record this value on the blank line in the top left box of your data table.

Lenses and Images *continued*

Forming Images

5. Set up your equipment.
   - Look at the illustration.
   - Set up your equipment as illustrated in the figure.

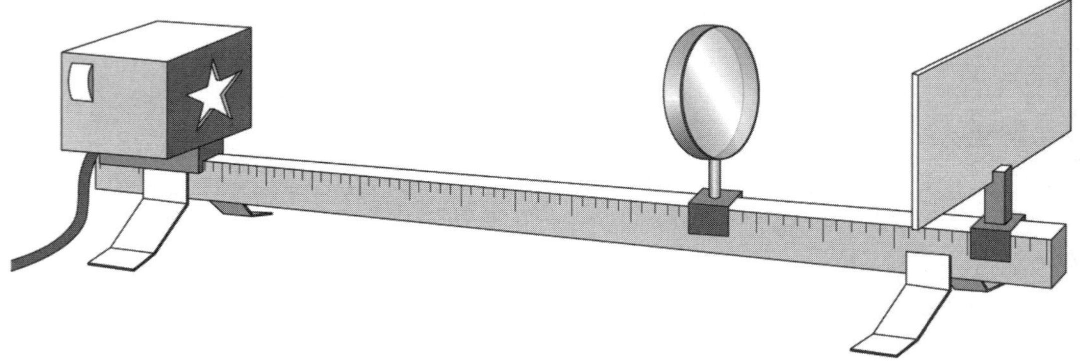

   - Look at the focal length you marked in your data table.
   - Place the lens more than twice the focal length from the light box.

6. Trial:
   - You may use either the filament of the light bulb or a cut-out shape in the light box as your object.
   - Move the screen along the meterstick until a clear image forms.
   - Use centimeters to measure the distance from the light to the lens.
   - This measurement is d_o, the object distance.
   - Record the distance the Trial 1 row of the "Object distance" column on the data table.
   - Use centimeters to measure the distance from the lens to the screen.
   - This measurement is d_i, the image distance.
   - Record the distance in the Trial 1 row of the "Image distance" column in your data table.
   - Use millimeters to measure the height of the object.
   - Record the measurement in the Trial 1 row of the "Size of object" column in your data table.
   - Use millimeters to measure the height of the image.
   - Record the measurement in the Trial 1 row of the "Size of image" column in your data table.

Lenses and Images *continued*

7. Trial 2:
   - Look at the focal length you recorded on the blank line in the top left box of your data table.
   - Place the lens exactly twice the focal length from the object.
   - Slide the screen along the stick until a clear image is formed, as you did in step 6.
   - Use centimeters to measure the distance from the light to the lens.
   - Record the distance the Trial 2 row of the "Object distance" column on the data table.
   - Use centimeters to measure the distance from the lens to the screen.
   - Record the distance in the Trial 2 row of the "Image distance" column in your data table.
   - Use millimeters to measure the height of the object.
   - Record the measurement in the Trial 2 row of the "Size of object" column in your data table.
   - Use millimeters to measure the height of the image.
   - Record the measurement in the Trial 2 row of the "Size of image" column in your data table.

8. Trial 3:
   - Look at the focal length you recorded on the blank line in the top left box of your data table.
   - Remember the distance you used as "twice the focal length" in Trial 2.
   - Place the lens at a distance from the object that is greater than the focal length but less than twice the focal length.
   - Move the screen along the meterstick until a clear image forms.
   - Use centimeters to measure the distance from the light to the lens.
   - Record the distance the Trial 3 row of the "Object distance" column on the data table.
   - Use centimeters to measure the distance from the lens to the screen.
   - Record the distance in the Trial 3 row of the "Image distance" column in your data table.
   - Use millimeters to measure the height of the object.
   - Record the measurement in the Trial 3 row of the "Size of object" column in your data table.
   - Use millimeters to measure the height of the image.
   - Record the measurement in the Trial 3 row of the "Size of image" column in your data table.

Lenses and Images *continued*

ANALYSIS

1. **Analyzing Data** Perform the calculations needed to complete your data table.
   - You my use a calculator to do the math.
   - Find $\dfrac{1}{d_o}$ for Trial 1:

   - Write the number in the Trial 1 row of the $\dfrac{1}{d_o}$ column of your data table.

   - Find $\dfrac{1}{d_o}$ for Trial 2:

   - Write the number in the Trial 2 row of the $\dfrac{1}{d_o}$ column of your data table.

   - Find $\dfrac{1}{d_o}$ for Trial 3:

   - Write the number in the Trial 3 row of the $\dfrac{1}{d_o}$ column of your data table.

Lenses and Images *continued*

- Find $\dfrac{1}{d_i}$ for Trial 1:

- Write the number in the Trial 1 row of the $\dfrac{1}{d_i}$ column of your data table.

- Find $\dfrac{1}{d_i}$ for Trial 2:

- Write the number in the Trial 2 row of the $\dfrac{1}{d_i}$ column of your data table.

- Find $\dfrac{1}{d_i}$ for Trial 3.

- Write the number in the Trial 3 row of the $\dfrac{1}{d_i}$ column of your data table.

Lenses and Images *continued*

- Find $\dfrac{1}{d_o} + \dfrac{1}{d_i}$ for Trial 1:

- Write the number in the Trial 1 row of the $\dfrac{1}{d_o} + \dfrac{1}{d_i}$ column of your data table.

- Find $\dfrac{1}{d_o} + \dfrac{1}{d_i}$ for Trial 2:

- Write the number in the Trial 2 row of the $\dfrac{1}{d_o} + \dfrac{1}{d_i}$ column of your data table.

- Find $\dfrac{1}{d_o} + \dfrac{1}{d_i}$ for Trial 3:

- Write the number in the Trial 3 row of the $\dfrac{1}{d_o} + \dfrac{1}{d_i}$ column of your data table.

Lenses and Images *continued*

- Look at the focal length of your lens you recorded on the blank line in the top left box of your data table.

- Find $\dfrac{1}{f}$.

____

- Record this number in the column $\dfrac{1}{f}$ for all three trials.

2. **Analyzing Data** Compare your data for $\dfrac{1}{d_o} + \dfrac{1}{d_i}$ with your data for $\dfrac{1}{f}$ for all three trials.

- Look at your data table. Are the measurements for $\dfrac{1}{d_o} + \dfrac{1}{d_i}$ more than, less than, or about the same as $\dfrac{1}{f}$ for all three trials?

____

____

____

COMMUNICATING RESULTS

3. **Drawing Conclusions** If the object distance is greater than the image distance, how will the size of the image compare with the size of the object?
   - The object distance was greater than the image distance in all three trials.
   - In Trial 1, was the size of the image larger or smaller than the size of the object?

____

   - In Trial 2, was the size of the image larger or smaller than the size of the object?

____

   - In Trial 3, was the size of the image larger or smaller than the size of the object?

____

Lenses and Images *continued*

APPLICATION

Look at the lens equations shown in step 1.

- Does the lens that you tested conform to the lens equations for image formation?

- Suppose you made a new model camera using this lens. What would the minimum length of the camera have to be? (**Hint:** Besides the focal length of the lens, what other measurement would you need to account for?)

Application Lab   **DATASHEET B**

Lenses and Images

As an optical engineer for a camera company, you have been given a lens for which your job is to figure out the focal length. Based on the specifications you obtain by doing an experiment, a new model of camera will be designed that uses that lens.

WHAT YOU'LL DO

Observe images formed by a convex lens.

Measure the distance of objects and images from the lens.

Analyze your results to determine the focal length of the lens.

WHAT YOU'LL NEED

- cardboard screen, 10 cm x 20 cm
- convex lens, 10 cm to 15 cm focal length
- lens holder
- light box with light bulb

- meterstick
- ruler, metric
- screen holder
- supports for meterstick

PROCEDURE

Preparing for Your Experiment

1. The shape of a lens determines the size, position, and types of images that it may form. When parallel rays of light from a distant object pass through a converging lens, they come together to form an image at a point called the *focal point*. The distance from this point to the lens is called the *focal length*. In this experiment, you will find the focal length of a lens. Then, verify this value by forming images, measuring distances, and using the lens formula,

$$\frac{1}{d_o} + \frac{1}{d_i} = \frac{1}{f}$$

where d_o = *object distance,*

d_i = *image distance,* and

f = *focal length*

Lenses and Images *continued*

2. Use the following table to record your data.

DATA TABLE: OBJECTS, LENSES, AND IMAGES

Focal length of lens, f: _____ cm	Object distance, d_o (cm)	Image distance, d_i (cm)	$\dfrac{1}{d_o}$	$\dfrac{1}{d_i}$	$\dfrac{1}{d_o}+\dfrac{1}{d_i}$	$\dfrac{1}{f}$	Size of object (mm)	Size of image (mm)
Trial 1								
Trial 2								
Trial 3								

3. Set up the equipment as illustrated in the figure below. Make sure the lens and screen are securely fastened to the meterstick.

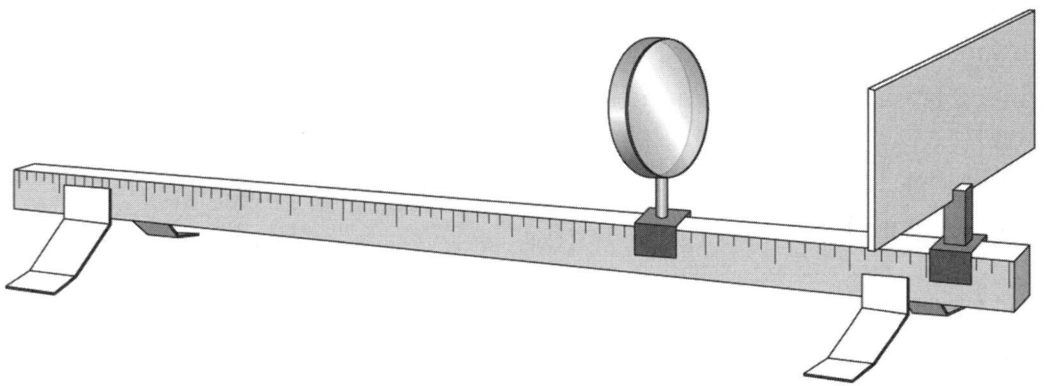

Determining Focal Length

4. Stand about 1 m from a window, and point the meterstick at a tree, parked car, or similar object. Slide the screen holder along the meterstick until a clear image of the distant object forms on the screen. Measure the distance between the lens and the screen in centimeters. This distance is very close to the focal length of the lens that you are using. Record this value at the top of your data table.

Forming Images

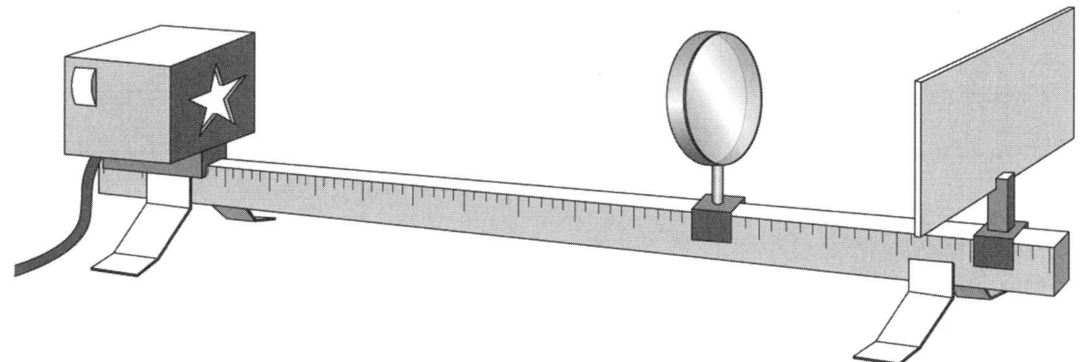

5. Set up the equipment as illustrated in the figure. Place the lens more than twice the focal length from the light box.

6. Move the screen along the meterstick until a clear image forms. Record the distance from the light to the lens, d_o, and the distance from the lens to the screen, d_i, in centimeters as Trial 1 in your data table. Also, record the height of the object and of the image in millimeters. The object in this case may be either the filament of the light bulb or a cut-out shape in the light box.

7. For Trial 2, place the lens exactly twice the focal length from the object. Slide the screen along the stick until a clear image is formed, as in step 6. Record the distances from the screen and the sizes of the object and image as you did in step 6.

8. For Trial 3, place the lens at a distance from the object that is greater than the focal length but less than twice the focal length. Adjust the screen, and record the measurements as you did in step 7.

ANALYSIS

1. **Analyzing Data** In the space below, perform the necessary calculations to complete your data table.

2. **Analyzing Data** How does $\dfrac{1}{d_o} + \dfrac{1}{d_i}$ compare with $\dfrac{1}{f}$ in each of the three trials?

____

____

____

COMMUNICATING RESULTS

3. **Drawing Conclusions** If the object distance is greater than the image distance, how will the size of the image compare with the size of the object?

____

____

____

APPLICATION

Does the lens that you tested conform to the lens equations for image formation? If a camera that contained this lens was made, what would the minimum length of the camera have to be? Explain your answer.

Application Lab | **DATASHEET C**

Lenses and Images

As an optical engineer for a camera company, you have been given a lens for which your job is to figure out the focal length. Based on the specifications you obtain by doing an experiment, a new model of camera will be designed that uses that lens.

WHAT YOU'LL DO

Observe images formed by a convex lens.

Measure the distance of objects and images from the lens.

Analyze your results to determine the focal length of the lens.

WHAT YOU'LL NEED

- cardboard screen, 10 cm x 20 cm
- convex lens, 10 cm to 15 cm focal length
- lens holder
- light box with light bulb
- meterstick
- ruler, metric
- screen holder
- supports for meterstick

PROCEDURE

Preparing for Your Experiment

1. The shape of a lens determines the size, position, and types of images that it may form. When parallel rays of light from a distant object pass through a converging lens, they come together to form an image at a point called the *focal point*. The distance from this point to the lens is called the *focal length*. In this experiment, you will find the focal length of a lens. Then, verify this value by forming images, measuring distances, and using the lens formula,

$$\frac{1}{d_o} + \frac{1}{d_i} = \frac{1}{f}$$

where d_o = *object distance,*

d_i = *image distance,* and

f = *focal length*

2. Read the steps of the lab and note the information you will record. Then, on a clean sheet of paper, make a data table.

3. Set up the equipment as illustrated in the figure below. Make sure the lens and screen are securely fastened to the meterstick.

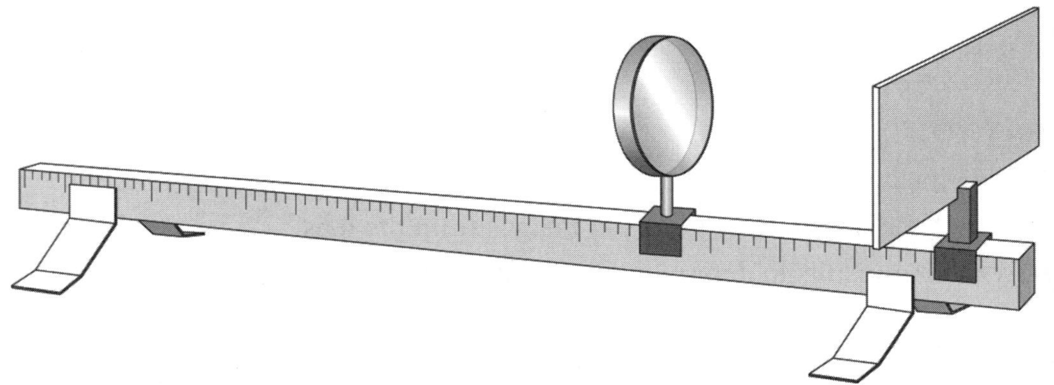

Determining Focal Length

4. Stand about 1 m from a window, and point the meterstick at a tree, parked car, or similar object. Slide the screen holder along the meterstick until a clear image of the object forms on the screen. Measure the distance between the lens and the screen in centimeters. This distance is very close to the focal length of the lens that you are using. Record this value in your data table.

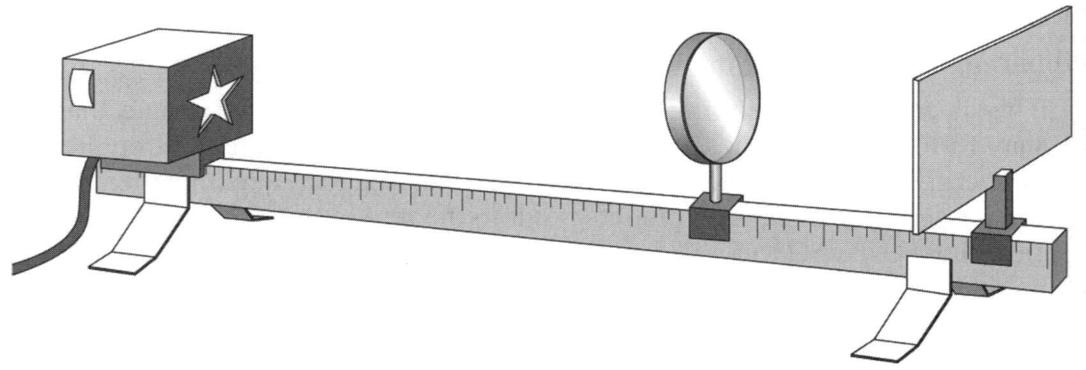

Forming Images

5. Set up the equipment as illustrated in the figure. Place the lens more than twice the focal length from the light box.

6. Move the screen along the meterstick until a clear image forms. Record in centimeters the distance from the light to the lens for Trial 1 in your data table. Also, record the height of the object and the image in millimeters in separate columns in your table.

7. For Trial 2, place the lens exactly twice the focal length from the object. Slide the screen along the stick until a clear image is formed, as in step 6. Record distances and sizes in your data table, as you did for Trial 1.

Lenses and Images *continued*

8. For Trial 3, place the lens at a distance from the object that is greater than the focal length but less than twice the focal length. Adjust the screen, and record the measurements as you did for Trials 1 and 1.

ANALYSIS

1. **Analyzing Data** Complete your data table. Using the focal length you found in step 4, record the focal length of the lens for all three trials. Calculate the object distance, and image distance for all three trials, and record the distances in your data table. In a separate column, calculate the focal length for all three trials, using the formula in step 1.

2. **Analyzing Data** How does $\dfrac{1}{d_o} + \dfrac{1}{d_i}$ compare with $\dfrac{1}{f}$ in each of the three trials?

____

____

____

COMMUNICATING RESULTS

3. **Drawing Conclusions** Explain how the sizes and distances of the object and image relate for all three trials.

____

____

____

APPLICATION

Does the lens that you tested conform to the lens equations for image formation? If a camera that contained this lens was made, what would the minimum length of the camera have to be? Explain your answer.

Skills Practice Lab   **OBSERVATION**

Mirror Images

Teacher Notes

TIME REQUIRED
One lab period

SKILLS ACQUIRED
Classifying
Collecting data
Constructing models
Identifying/recognizing patterns
Measuring
Organizing and analyzing data

THE SCIENTIFIC METHOD

Make observations In the procedure steps, students will make observations regarding virtual images, flat mirrors, and curved mirrors.

Analyze the results In Analysis questions 1–7, students will analyze the data and results from their experiments.

Draw conclusions In Conclusions questions 1–4, students use their results to draw conclusions about mirror images.

MATERIALS

If a reverse eye chart is unavailable, use the normal eye chart for both parts of the activity.

Any small object can substitute for the T-pin and pencil eraser.

Mount all mirrors in cardboard to protect students from sharp edges. Use a craft knife to cut cardboard squares to make a frame for each mirror, and tape the mirror securely into the frame.

For the curved mirrors, you may either use separate concave and convex mirrors, or a double-sided curved mirror (one concave side and one convex side).

SAFETY CAUTIONS

Discuss all safety symbols and cautions with students.

Secure mirrors to the wall with strong tape and to the table with clamps or strong tape. The mirror taped to the wall should be lightweight and no larger than 10 cm by 15 cm.

The pencil eraser used on the T-pin should be new and not worn. The purpose of putting the eraser on the tip of the pin is to protect the student's eyes (it also makes the pin easier to see). Make sure that students wear safety goggles. It is possible to substitute another small object, such as a larger eraser or the cap of a pen, for the T-pin.

MISCONCEPTION ALERT

Students may confuse the terms *concave* and *convex.* You may point out to them that a concave mirror is recessed inward, like a cave.

TIPS AND TRICKS

Students can work alone or in groups of two or more.

It is more efficient to set up the lab in stations so that groups can rotate from station to station.

In step 4, make sure that the mirror is small and lightweight. A mirror that is too heavy will fall and break. Tape all sides of the mirror to the wall.

In step 8, secure mounted mirrors to the table by using clamps or strong tape to prevent broken-glass hazard.

In step 13, the angle between the incoming beam and the nearest perpendicular line should be approximately equal to the angle between the outgoing beam and the nearest perpendicular line.

In step 18, during the bench test at close range (20 cm), objects appear enlarged and upright; at middle range (30 cm), objects appear reduced and upright; at long range (65 cm), objects appear greatly reduced and upright.

In step 21, during the bench test at close range (20 cm), objects appear enlarged and upright; at middle range (30 cm), objects appear greatly enlarged and upright; at long range (65 cm), objects appear reduced and inverted.

TECHNIQUES TO DEMONSTRATE

Show students how to properly secure mirrors to the table by using mirror supports and to the wall by using strong tape.

Skills Practice Lab

Mirror Images

Introduction

Mirrors form images by reflecting light coming from objects. When you view an image in a mirror with your eyes, you are seeing a virtual image, which usually appears to exist somewhere behind the mirror. Depending on the shape of the surface of the mirror, this mirror image may be reversed left to right, turned upside down, or magnified to look larger or smaller. In this lab, you will observe images in a flat mirror, a convex mirror (curved outward), and a concave mirror (curved inward). You will also measure various distances and angles to learn more about the properties of the images formed by these different types of mirrors.

OBJECTIVES

Describe the images of objects in various kinds of mirrors.

Draw diagrams showing paths of light rays reflected from mirrors.

Compare light path lengths and angles of incidence and reflection.

MATERIALS

- eye charts, both normal and reverse
- meterstick
- mirror, curved
- mirror, small and flat
- mirror supports
- paper
- pencil
- protractor
- ruler or straightedge
- tape
- T-pin with pencil eraser
- white paper

SAFETY

PROCEDURE

Virtual images

1. Secure the normal eye chart to the wall by using strong tape.

2. Choose any line on the chart, and step back until the line can no longer be read clearly. Use masking tape to mark the position on the floor where you are standing. Label this position "reading point."

3. Measure the distance from the eye chart to the reading point with a meterstick. Record this distance using the appropriate SI units (meters). Also, record the number of the line that you were trying to read.

Mirror Images *continued*

4. Secure a small, flat mirror against the wall at chest level by using strong tape.

5. Place the back of the reverse eye chart against your chest. Position the chart so that the line that you read appears in the mirror. Step back from the mirror, and hold the eye chart against your chest until the image of this line is barely readable.

6. Use masking tape to mark the position on the floor where you are standing. Label this position "new point."

7. Measure the distance from the eye chart to the new point. Record this distance using the appropriate SI units.

Flat mirrors

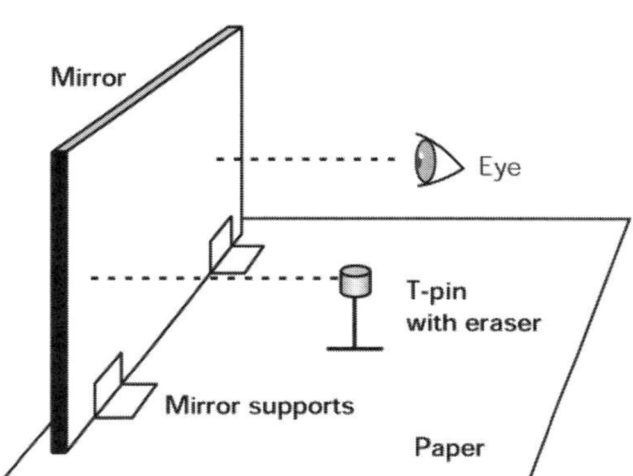

8. Using two mirror supports, vertically stand one flat mirror on a table, away from the edge, as shown in the figure above. Place a sheet of white paper on the tabletop so that the front of the mirror faces the paper. Tape the paper and mirror supports to the table so that they do not slide.

9. Remove the eraser from a pencil. Secure the eraser on the tip of the T-pin to cover the point. Using tape, carefully secure the T-pin on the tabletop, with the T side down, in front of the mirror. The point with the eraser should be pointing up.

10. Wearing a pair of safety goggles, move your head to one side of the pin. Close one eye, and place your open eye at the level of the tabletop. Observe the image of the pin in the mirror.

11. Use a ruler to draw a straight line on the paper from the image of the pin in the mirror to the position of your eye. Label this line "outgoing beam." Use a ruler to draw a straight line from the object to the mirror's surface. Label this line "incoming beam." Both lines should meet at the same point on the mirror's surface.

12. Draw a line on the paper from the position of your eye perpendicular to the mirror's surface. Draw a line from the object perpendicular to the mirror's surface. Both lines should be parallel to each other. These lines will form angles with the lines you drew in step 11.

13. Measure the angle between the line labeled "outgoing beam" and the nearest perpendicular line. Measure the angle between the line labeled "incoming beam" and the nearest perpendicular line. Record these angles using units of degrees.

____

____

14. Move your eye to a new position. Repeat steps 10–13.

____

____

15. Move your eye to a third position. Repeat steps 10–13.

____

____

Curved mirrors

Mirror
that curves
outward

Mirror
that curves
inward

16. Obtain a curved mirror. Use one mirror support to hold the mirror upright on the bench. Place the mirror so that you are facing the side that curves outward (the convex side).

17. Place an object at various distances from the mirror. Look at the image of the object in the mirror.

18. Observe and record in Table 1 how the image appears. Include the object's position (close to the mirror or far from the mirror), the size of the image (enlarged or small), and the orientation of the image (upright or upside down).

TABLE 1 CONVEX MIRROR

Distance from Mirror (cm)	Size of Image	Orientation of Image

19. Turn the mirror around so that you are facing the side that curves inward (the concave side).

20. Place an object at various distances from the mirror. Look at the image of the object in the mirror.

21. Observe and record in Table 2 how the image appears. Include the object's position (close to the mirror or far from the mirror), the size of the image (enlarged or small), and the orientation of the image (upright or upside down).

TABLE 2 CONVEX MIRROR

Distance from Mirror (cm)	Size of Image	Orientation of Image

ANALYSIS

Virtual images

1. **Describing events** Describe the image of the reverse eye chart you saw on the surface of the mirror. Compare it with the appearance of the normal eye chart.

2. **Examining data** What distance did you measure between the mirror and the reverse eye chart?

3. **Examining data** What distance did you measure between the starting point and the eye chart on the wall?

Flat mirrors

4. **Organizing data** In your notebook, draw the experimental setup as viewed from above. Include the lines and angles for each trial.

Mirror Images *continued*

Curved mirrors

5. **Describing events** How did the image appear when the object was in front of the convex mirror?

____

____

6. **Describing events** How did the image appear when the object was close to the concave mirror?

____

____

7. **Describing events** How did the image appear when the object was far away from the concave mirror?

____

____

CONCLUSIONS

1. **Drawing conclusions** Compare your answers to Analysis questions 2 and 3 above. What is the relationship between the distances?

____

____

____

2. **Drawing conclusions** Compare the two angles measured in step 13 for each position. What is the relationship between the angles?

____

____

3. **Classifying** Which kinds of mirrors used in this experiment are capable of producing images that are reversed left to right?

____

____

4. **Classifying** Which kinds of mirrors used in this experiment are capable of producing images that are inverted (upside down)?

____

____

EXTENSIONS

1. **Research and communications** Many side mirrors on automobiles have a warning label, "Objects in mirror are closer than they appear." What kind of mirror do you think these are? What might be the advantage of using this kind of mirror for this application? Research these types of mirrors on the Internet or in a library to check your answers.

____

____

____

2. **Building models** If you were designing a house of mirrors for a carnival, what kind of mirror would you use to create an upside down image? Would the person looking in the mirror have to be close to the mirror or far away from it to see his or her image upside down?

____

____

____

____

____

____

____

Choosing a Pair of Sunglasses

Teacher Notes

TIME REQUIRED
One lab period

SKILLS ACQUIRED
Collecting data
Experimenting
Identifying patterns
Interpreting
Measuring
Organizing and analyzing data
Constructing models
Communicating

RATING

Easy ←——1——2——3——4——→ Hard

Teacher Prep–1
Student Set-Up–3
Concept Level–4
Clean Up–1

THE SCIENTIFIC METHOD

Make observations In the Procedure, students measure and record the intensity of light transmitted through different pairs of sunglasses.

Analyze the results In Analysis questions 1 and 2, students calculate different intensities of light and graph their results.

Draw conclusions In Conclusions questions 3, 4, and 5, students compare the different sunglasses.

MATERIALS

If possible, have a light sensor, a CBL, and a graphing calculator connected for each group ahead of time. The calculators should be loaded with the Vernier PHYSCI program.

Arrange for students to bring in sunglasses on the day of the lab. It might be a good idea to have darkened glass or colored cellophane on hand for those that do not have sunglasses or forget to bring them. Have a piece of polarized glass on hand for each group in case students do not have sunglasses with polarized lenses.

The desk lamp should be equipped with a 60 W light bulb.

Give each group a hot pad or a padded glove in case they need to adjust the lamp while it is hot.

If a piece of glass or aluminum foil is not available, a flat tray of water will also work.

Choosing a Pair of Sunglasses *continued*

SAFETY CAUTIONS

Discuss all safety symbols and cautions with students.

Students should wear safety goggles and gloves while performing this experiment.

If students are working in the laboratory, they should also wear laboratory aprons.

Make sure students do not touch the light source or look directly at it. Remind students to allow ample time for the lamp to cool before they handle it.

Make sure that students wash their hands when they are done with the experiment.

TECHNIQUES TO DEMONSTRATE

Remind students to check the intensity of the reflected light before they test each pair of sunglasses.

TIPS AND TRICKS

Having two to four students in each lab group works best.

In steps 9 and 10, be sure that students position the sunglasses so that the light passes through one of the lenses and not between the lenses through the nose piece.

In step 12, do not let students look directly at the light bulb or at the sun regardless of whether they are wearing sunglasses or not.

Skills Practice Lab **CBL PROBEWARE**

Choosing a Pair of Sunglasses

Introduction

Very intense reflected sunlight, or glare, can sometimes make it difficult for you to see. Wearing sunglasses is one way to protect your eyes from such blinding light. There are many different sunglasses to choose from, so how do you decide which pair to buy? Imagine that a consumer magazine has hired you to compare several different pairs of sunglasses for their ability to block reflected light. You must also determine how well you can see while wearing each pair.

OBJECTIVES

Measure the intensity of light that is reflected from a smooth, flat surface.

Compare several pairs of sunglasses for their ability to block reflected light while still allowing you to see well.

Evaluate your data to determine which pair of sunglasses you would recommend that people wear.

MATERIALS

- light sensor
- CBL
- TI graphing calculator
- black link cable
- desk lamp
- piece of glass or aluminum foil

- ring stand
- clamp
- several different pairs of sunglasses, some having polarized lenses

SAFETY

FINDING OUT MORE INFORMATION

If light is reflected from a rough surface, the light is reflected in many different directions. But light is reflected all in the same direction from a smooth surface, such as water, glass, or a flat piece of metal. The result is an unpleasant glare if the light is directed toward your eyes.

An effective pair of sunglasses is able to reduce the intensity of reflected light while still letting enough light pass through the lenses for you to see well. Sunglass manufacturers either add colors to the lenses, coat the lenses with reflective paints, or polarize the lenses (meaning that they make the lenses so that only light traveling in a certain direction can pass through).

COMING UP WITH A PLAN

The first thing you must do is simulate reflected sunlight. You can do this by setting up a desk lamp so that it reflects off glass or some other smooth surface. Then you can measure the intensity of the reflected light using a light sensor.

Next you will place the sunglasses between the reflected light and the sensor. This will allow you to measure how much of the reflected light is able to pass through the lenses of each pair of sunglasses.

You can determine whether the lenses are polarized by comparing how much light passes through them when they are positioned straight and then rotated. If the lenses aren't polarized, the intensity of the light passing through them will be about the same whether the lenses are rotated or not. If the lenses are polarized, the intensity of the light passing through them will be different when they are rotated.

PROCEDURE

1. Use the table below to record your data.

SUNGLASSES DATA

Light intensity reading without sunglasses	Brand/style of sunglasses	Light intensity reading with sunglasses horizontal	Light intensity reading with sunglasses vertical	Maximum difference in light intensity readings

Setting Up the CBL System

2. Plug the light sensor into the Channel 1 input of the CBL. Connect the CBL to the graphing calculator by plugging the black link cable into the base of each unit.

3. Turn on both the CBL and the calculator. Press PRGM on the calculator, and select the PHYSCI program.

4. Go to the MAIN MENU, and select SET UP PROBES. Enter "1" as the number of probes. Select LIGHT from the SELECT PROBE menu. Enter "1" as the channel number.

5. From the MAIN MENU, select COLLECT DATA. Select MONITOR INPUT from the DATA COLLECTION menu. Light intensity readings will be displayed on the calculator.

| **Choosing a Pair of Sunglasses** *continued*

Reflecting Light

6. Set up the desk lamp, reflective surface, and light sensor as shown in **Figure 1.** **Do not place a pair of sunglasses in front of the light sensor yet.**

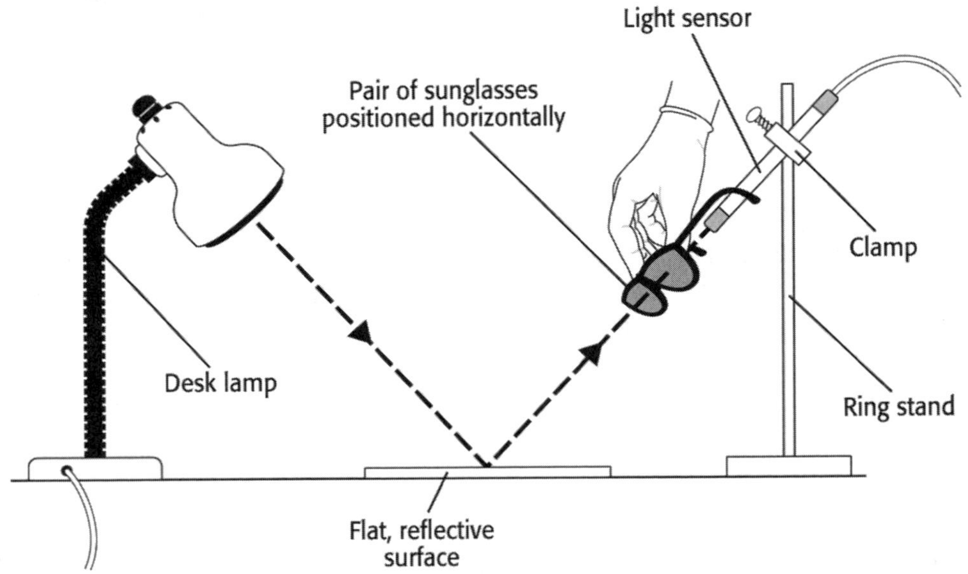

Figure 1

7. Adjust the lamp and light sensor so that the light sensor receives the maximum reflected light. Secure the light sensor with a clamp. You may also want to secure the desk lamp.

8. Without any sunglasses in front of the light sensor, record the intensity of the reflected light in the data table.

Testing the Sunglasses

9. Hold a pair of sunglasses directly in front of the light sensor. Orient the glasses horizontally, as shown in **Figure 1** above. Record the intensity of the light in the data table.

10. Rotate the sunglasses 90 degrees, as shown in **Figure 2.** Record the intensity of the light in the data table.

Choosing a Pair of Sunglasses *continued*

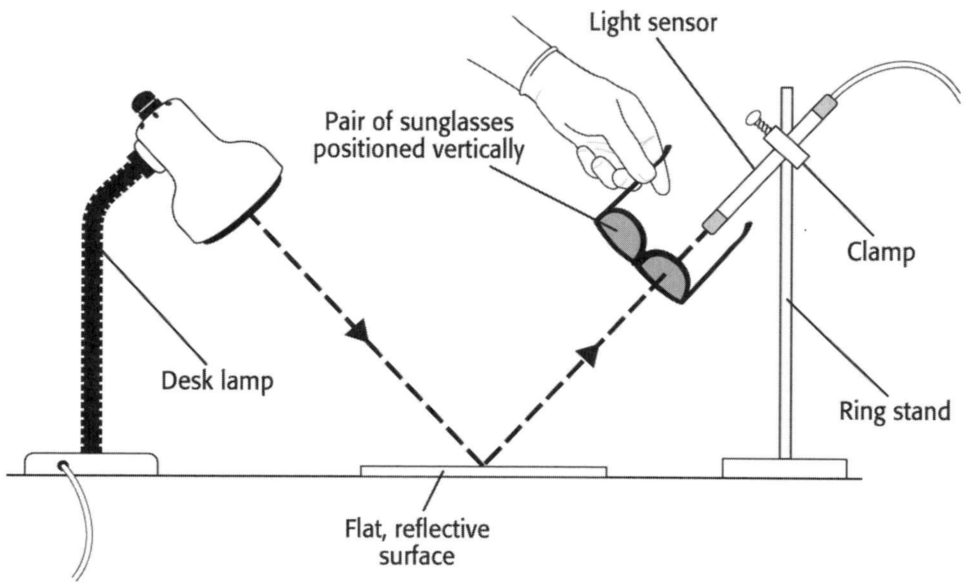

Figure 2

11. Note what color the lenses are and any other important details about the sunglasses.

Brand X: ____

Brand Y: ____

Brand Z: ____

12. Put the sunglasses on, and look out a window at some reflected glare. (Do not look directly at the sun regardless of whether you are wearing sunglasses or not.) Then try reading some fine print in your textbook. Note how well you can see.

Brand X: ____

Brand Y: ____

Brand Z: ____

13. Repeat steps 8–12 for each pair of sunglasses. Be sure to check the light intensity without sunglasses first each time to make sure the lighting conditions haven't changed.

14. When you have finished, press "+" on the calculator. Then turn off the light, put away all of your materials, and clean your work area. Be sure to wash your hands thoroughly before you leave.

Choosing a Pair of Sunglasses *continued*

ANALYSIS

1. Calculate the maximum difference in light intensity for each pair of sunglasses by subtracting the smaller of the two measurements made with sunglasses from the measurement made without any sunglasses. Record your answers in the data table.

____

____

____

____

2. Plot your data on the bar graph below. For each pair of sunglasses, plot the light intensity without any sunglasses (None), the light intensity with the sunglasses positioned horizontally (H), and the light intensity with the sunglasses positioned vertically (V).

Comparing Sunglasses

Light intensity reading (y-axis: 0, 0.05, 0.10, 0.15, 0.20, 0.25, 0.30)

None H V — Pair 1 | None H V — Pair 2 | None H V — Pair 3

CONCLUSIONS

3. Which pairs of sunglasses have polarized lenses? Could you read fine print very well while wearing these sunglasses? Explain why or why not.

____

____

____

____

Choosing a Pair of Sunglasses *continued*

4. Which sunglasses were able to reduce the reflected light intensity the most? Could you read fine print very well while wearing these sunglasses? Explain why or why not.

____

____

____

____

5. Of the sunglasses you tested, choose the pair that you think is the best overall. Explain your reasoning.

____

____

____

____

6. In this experiment, you tested only two of the many factors that make sunglasses effective. What are some other important factors in sunglass design?

____

____

____

____

____

EXTENSION

1. Perform a similar experiment to compare the intensity of light that is reflected from various flat surfaces. Try mirrors, water, bricks, light and dark paper, CD jewel cases, and anything else you have available.

Answer Key

Concept Review

SECTION: SOUND

1. The speed of sound depends on how often the particles of the medium collide with one another. At higher temperatures, the particles move faster and collide more often.
2. The loudness of a sound depends partly on the energy contained in the sound waves and on intensity, which depends on the waves' amplitude. The higher the intensity of a sound is, the louder the sound will seem.
3. The pitch produced by a stringed instrument can be changed by moving your finger to a different position on the string, thereby increasing or decreasing the length of the string. A shorter length of string vibrates at a higher frequency, and a longer length of string vibrates at a lower frequency.
4. Infrasound is any sound below the range of human hearing, and ultrasound is any sound with a frequency above the range of human hearing. Infrasound is used by animals to communicate. Ultrasound is used to see inside the human body.
5. Sound waves in the air vibrate the eardrum. These vibrations pass from the eardrum to the bones of the middle ear—the hammer, anvil, and stirrup. The stirrup strikes a membrane at the opening of the inner ear, which sends waves through the cochlea.
6. a. 3,850 m
   $d = vt = (1{,}540 \text{ m/s})(2.50 \text{ s}) = 3{,}850 \text{ m}$
   b. 1,920 m
   $d = (3{,}850 \text{ m})/2 = 1{,}920 \text{ m}$
   (the depth of the water is half the distance traveled by the sound pulse)

SECTION: THE NATURE OF LIGHT

1. Light can be modeled either as a wave or as a stream of particles. Depending on the situation, light seems to behave either like a wave or like a particle.

2. a. wave model
   b. particle model
   c. particle model
3. a. radio waves
   b. gamma rays
   c. gamma rays
   d. radio waves
4. Light spreads out in spherical wave fronts. As light spreads out from the source, the number of photons passing through a given area on a sphere decreases.
5. radio waves, microwaves, infrared light, visible light, ultraviolet light, X rays, gamma rays
6. X rays are used to produce images of bones and other structures. Gamma rays can be used to treat cancer by killing diseased cells.

SECTION: REFLECTION AND COLOR

1. Light rays striking a rough surface are reflected in many directions, while light rays striking a smooth surface are all reflected in the same direction. Rough surfaces cause diffuse reflection because light rays are reflected in many directions when they hit the uneven surface (the normal to the surface is not always pointing in the same direction).
2. Check students' drawings for accuracy. The angle of incidence and the angle of reflection should appear to be roughly 30°, measured from the normal (perpendicular). The light rays should show directional arrows, and the angle of incidence should lie on the side of the normal where light rays are approaching.
3. The eye would perceive an image created by a concave mirror as larger than the image really is. An image created by a convex mirror appears smaller than the image really is.
4. Check students' drawings for accuracy.
5. a. yellow
   b. yellow
   c. black, because blue light contains no yellow light

SECTION: REFRACTION, LENSES, AND PRISMS

1. a. toward the normal
   b. away from the normal
2. You see a mirage when light is refracted by hot air just above the ground. Your brain may interpret the image of the sky coming from the direction of the ground as a reflection caused by water.
3. red, orange, yellow, green, blue, violet
4. A converging lens creates either virtual or real images. Incoming light rays are bent inward by a converging lens.
5. a. The cornea refracts light.
   b. The pupil allows light to pass through to the lens.
   c. The lens refracts light onto the retina at the back of the eye.
6. White light is made up of light of different colors. Because light of different colors travels at different speeds in a medium, a prism refracts each color through a different angle. As a result, white light is separated out, or dispersed, into a spectrum of colors.

Science Skills
ANGLES AND DEGREES

1. 30°
2. 50°
3. yes
4. 360°

THE AREA OF A CIRCLE

1. 491 cm^2
2. 200 m^2 (2 significant figures)

EQUATIONS INVOLVING A CONSTANT

1. 4.0×10^{-7} m
2. 170 kg • m/s^2, or 170 N

Cross-Disciplinary
INTEGRATING SPACE SCIENCE: THE REFRACTING TELESCOPE AT YERKES

1. The 40-inch telescope at Yerkes is the largest refracting telescope in the world.
2. A movable floor is necessary because the telescope pivots around the middle of its tube; as one end is moved to observe different parts of the sky, the other end (which contains the eyepiece) also moves. As a result, the eyepiece is not always at the same height.

3. Stars move so slowly that smaller time periods do not reveal their motion.

REAL WORLD APPLICATIONS: HOW DOES SUNSCREEN WORK?

1. Sunburns occur when ultraviolet radiation causes the expansion of small vessels in the skin that contain red and white blood cells. Suntans occur when melanin is released to the outer surface of the skin as a protection against ultraviolet radiation.
2. The pigments in sunscreen and melanin both serve the same function: absorbing ultraviolet radiation so that it does not penetrate the skin.
3. 20 SPF sunscreen absorbs 19/20 of the UV rays, allowing only 1/20 to penetrate the skin.

Pretest

1. a
2. b
3. c
4. d
5. Check students' drawings for accuracy. The transverse wave should approximate a sine wave; the longitudinal wave should show alternating areas of higher density and lower density. The wavelength on each type of wave should mark the distance between two successive identical parts of the wave. On the transverse wave, make sure the amplitude marks only half the vertical distance from crest to trough. Amplitude cannot be indicated directly on the longitudinal wave.
6. The frequency of a wave can be determined by counting the number of wavelengths that go past a point in a certain amount of time. The frequency in hertz (Hz) is equal to the number of cycles (full wavelengths) divided by the time in seconds.

Quizzes
SECTION: SOUND

1. c	6. c
2. b	7. b
3. b	8. c
4. a	9. b
5. d	10. a

SECTION: THE NATURE OF LIGHT

1. d	6. e
2. a	7. a
3. a	8. f
4. c	9. d
5. b	10. c

SECTION: REFLECTION AND COLOR

1. a	6. c
2. c	7. a
3. c	8. b
4. a	9. a
5. b	10. d

SECTION: REFRACTION, LENSES, AND PRISMS

1. b	6. b
2. d	7. f
3. c	8. c
4. a	9. a
5. e	10. d

Chapter Tests

TEST A (5 POINTS PER ITEM)

1. b
2. a
3. a
4. d
5. a
6. a
7. a
8. d
9. b
10. a
11. intensity or amplitude
12. sonar
13. hammer; anvil; stirrup
14. photons
15. electromagnetic spectrum
16. normal
17. virtual
18. rainbow
19. normal
20. Scientific theories are successful to the extent that they explain observations. Some observations are best explained by a wave model of light. Other observations are best explained by a particle model. Therefore, both models are currently accepted as correct, and light is considered to have a dual nature.

TEST B (5 POINTS PER ITEM)

1. d
2. a
3. c
4. c
5. b
6. d
7. b
8. d
9. d
10. d
11. temperature
12. basilar membrane (or cochlea)
13. ultrasound
14. particles
15. electromagnetic spectrum
16. diffuse reflection
17. convex
18. reflects; absorbs
19. refract
20. Answers may vary and should include two of the following: X rays, with high energy and short wavelengths, are used to produce images of the body's interior; infrared, with wavelengths longer than red light, can be used to keep food warm; microwaves, with still longer wavelengths, can be used for cooking and for communication; radio waves, with the longest wavelengths, can be used in communication and radar.

Standardized Test Practice

1. C
2. B
3. D
4. A

Inquiry Lab: Colors in White Light

1. When I looked at the incandescent light I saw a continuous spectrum of colors on the sides of the spectroscope tube. When I looked at the fluorescent light I saw a spectrum of colors, but some bands were bright and others were faint.
2. White light is made up of many colors of light.
3. I would expect to see more colors than just yellow.

Quick Lab: Sound in Different Mediums

ANALYSIS

1. The sound that I heard through the string was louder. It was a ringing, bell-like sound.
2. Sound travels better through the string. The sound was louder.
3. The sound waves that reached my ear through the string were more intense than the sound waves that traveled through air. The intensity was due to sound waves traveling faster through solids than gases.

Quick Lab: Frequency and Pitch

1. Sample answer: Yes, I can hear the sound.
2. When less of the ruler vibrates, the ruler is vibrating at a higher frequency, so the sound has a higher pitch.

Inquiry Lab: How Can You Amplify the Sound of a Tuning Fork?

ANALYSIS

1. Sample answer: The objects, such as desktops, boxes, tabletops, and sheets of poster board, were made of hard, thin materials and had a relatively large surface area.
2. Yes, this works when using two tuning forks with the same pitch.
3. Sample answer: Yes, if I touch the base of a ringing tuning fork to a resonant object and then touch a tuning fork of the same frequency to the object, the second tuning fork starts vibrating.
4. They both ring with the same frequency, or pitch.

Quick Lab: Echoes and Distance

1. Sample answer: Yes, I heard an echo.
2. Sample answer: It took 1.2 s for the sound to return.
3. Sample answer:

$$d = vt = \frac{(340 \text{ m/s} \times 1.2 \text{ s})}{2} = 204 \text{ m}$$

Quick Lab: Curved Mirror

ANALYSIS

1. The front side of the spoon curves inward and is the concave mirror. The back side of the spoon curves outward and is a convex mirror.
2. The image in the concave part of the spoon was small and upside down, except when the pencil was very close to the spoon. The image was smaller than the pencil in the convex part of the spoon.
3. Ansers may vary. In the concave part of the spoon, the image of the pencil was larger than the pencil when the pencil was very close to the mirror. As the distance between the pencil and the concave mirror increased, the image inverted and became smaller than the actual pencil. For the convex mirror, the image was always right side up and smaller, and the image became smaller the farther the pencil was from the spoon.

Quick Lab: Filtering Light

1. Answers may vary depending on the filters used and the objects being viewed. Sample answer: When I looked through the yellow filter, the color of the object changed. Objects looked more yellow, and some objects looked darker.
2. Answers may vary depending on the filters used and the objects being viewed. Sample answer: When I looked through the yellow filter, the color of the object changed. Objects looked more yellow, and some objects looked darker.
3. Answers may vary. When I combined the yellow and the blue filter, the objects looked greener and some objects appeared to be darker. The colors of the objects were different than when I looked at them through a single filter because each filter absorbs certain wavelengths of light, so those wavelengths are subtracted out from what I see. Fewer colors pass through two combined filters than through one filter.
4. When I look through all fo the filters at once, the objects are very dark and have very little color.

Quick Lab: Water Prism

3. Answers may vary depending on quality of reflection. Sample answer: I see red, orange, yellow, green, blue, and violet.
4. When light travels through the water, the wavelengths of visible light travel at different speeds. Red light travels fastest and is bent the least, and violet travels the slowest and is bent the most. As a result, the colors are dispersed, or spread out.

Chapter Lab: Lens and Images

DATASHEETS A, B, & C
Analysis

1. Results may vary. See sample data table below.

Focal length of lens: 10.2 cm	Object distance, d_o (cm)	Image distance, d_i (cm)	$\dfrac{1}{d_o}$	$\dfrac{1}{d_i}$	$\dfrac{1}{d_o} + \dfrac{1}{d_i}$	$\dfrac{1}{f}$	Size of object (mm)	Size of image (mm)
Trial 1	22.0	18.2	0.045	0.055	0.100	0.100	19	16
Trial 2	20.4	21.6	0.049	0.046	0.095	0.095	19	18
Trial 3	16.5	28.0	0.061	0.036	0.097	0.097	19	32

2. The value of $\dfrac{1}{d_o} + \dfrac{1}{d_i}$ is generally very close to $\dfrac{1}{f}$.

Communicating Your Results

3. If the object distance is greater than the image distance, the image will be smaller than the object.

Application

Yes, the experiments showed that the lens forms images according to the standard lens equations. The minimum length of the camera would have to be the focal length of the lens, plus whatever additional space is needed for the camera housing and other componenents that would go behind the camera's image-capturing mechanism.

Observation: Mirror Images

PROCEDURE

3. Answers may vary depending on the line chosen and visual acuity of student.
7. Answers may vary.
13. Answers may vary. Each "incoming beam" should have the same angle measure as its corresponding "outgoing beam."
14. Answers may vary. Each "incoming beam" should have the same angle measure as its corresponding "outgoing beam."
15. Answers may vary. Each "incoming beam" should have the same angle measure as its corresponding "outgoing beam."

ANALYSIS

1. The image of the reverse eye chart seen in the mirror is identical to the appearance of the normal eye chart viewed directly (it is no longer reversed).
2. Answers may vary.
3. Answers may vary.
4. Diagrams should show that the incoming and outgoing angles are equal in each trial.
5. When an object is in front of a convex mirror, its image appears upright and smaller than the object.
6. When an object is close to a concave mirror, its image appears upright and larger than the object.
7. When an object is far from a concave mirror, its image appears inverted and smaller than the object.

CONCLUSIONS

1. The distance from the chart to the mirror is approximately one-half the distance from the student to the chart. Some students may realize that the distance between the image and the eye is actually the same in both cases.
2. The angles are equal.
3. all three types: flat, concave, and convex
4. Only the concave mirror can create an inverted image.

EXTENSIONS

1. Automobile side mirrors that have the warning label are convex mirrors. The primary advantage of using a convex mirror is that it provides a wider field of view.
2. You would use a concave mirror. The person would have to stand far from the mirror.

CBL Probeware: Choosing a Pair of Sunglasses

PROCEDURE

1.

Sunglasses Data

Light intensity reading without sunglasses	Brand/ style of sunglasses	Light intensity reading with sunglasses horizontal	Light intensity reading with sunglasses vertical	Maximum difference in light intensity readings
0.258	Brand X	0.165	0.163	0.095
0.248	Brand Y	0.056	0.058	0.192
0.257	Brand Z	0.078	0.112	0.179

11. Answers may vary.
12. Answers may vary.

ANALYSIS

1. Answers may vary. The calculation for Brand X is shown below. Answers for all three pairs of sunglasses are listed in the table above. Maximum difference in readings for Brand X = 0.258 – 0.163 = 0.095

2. Graphs may vary. For the sample data provided:

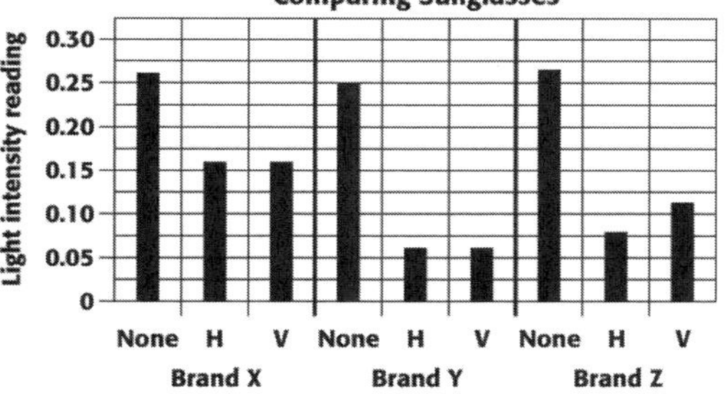

CONCLUSIONS

3. Answers may vary. For the sample data provided, only the Brand Z sunglasses were polarized. Students should be able to read fine print while wearing polarized sunglasses because the lenses are not dark enough to hinder reading.

4. Answers may vary. For the sample data provided, Brand Y sunglasses were able to reduce the light's intensity the most, followed by Brand Z and then Brand X.

5. Answers may vary. Based on the sample data, Brand Y sunglasses blocked the most light, but they did not allow for reading fine print as well as Brand Z. Therefore, Brand Z is the best pair of sunglasses overall.

6. Answers may vary. One factor that was not tested in this experiment that is also important in sunglass design is the effect of different kinds of lens coatings and tints. This experiment also did not test the ability of the sunglasses to filter ultraviolet (UV) light. Finally, this experiment did not take into account people's personal preferences for different styles of sunglasses.

EXTENSION

1. Students can either bring in items from home or test various items that are in the lab. Have students make a bar graph that compares the light each surface reflects.